THE SOLAR SYSTEM: THE SUN AND THE PLANETS

THE SOLAR SYSTEM
THE SUN AND THE PLANETS

THE SOLAR SYSTEM: THE SUN AND THE PLANETS

INDEX
 3 **THE SOLAR SYSTEM**
 5   Minor bodies of the solar system:
 7 SUN
10 COMETS
12 METEORITES
14 MERCURY
15 VENUS
17 MOON
20   Trip to the Moon
21 THE EARTH
26 MARS
30 **ASTEROID BELT**
32 •CERES
34 •VESTA
35 •PALLAS
36 **JUPITER**
37 •GANYMEDE
38 •CALLISTO
39 •IO
40 •EUROPA
42 **SATURN**
44 •TITAN
46 •RHEA
47 •IAPETUS
47 •ENCELADUS
50 •PHOEBE
51 **URANUS**
53 •TITANIA
54 •MIRANDA
55 •OBERON
55 **NEPTUNE**
56 •TRITON
58 •NEREID
59 **PLUTO**
61 •CHARON
62 **DWARF PLANETS BEYOND NEPTUNE**
62 •ERIS
63 •SEDNA
63 •MAKEMAKE
64 •HAUMEA
64 •QUAOAR
65 •GONGGONG
67 Acknowledgements

THE SOLAR SYSTEM

Pythagoras already said that the earth was a sphere, based on the observation of the shadow of eclipses; and in the 3rd century BC. **Aristarchus** was a supporter of the heliocentric model, but the geocentric model (planets and sun revolving around the Earth) was mostly accepted until **Nicholas Copernicus.**
Galileo discovers that there are satellites orbiting Jupiter, and is accused of heresy.
Johannes Kepler mathematically explains how the planets move around the sun. Later, **Isaac Newton** determined the laws of gravity.

The solar system appears 4.6 billion years ago from the collapse of a **cloud of stellar dust** that, due to the force of gravity, created a protoplanetary disk, from which the planets were formed.
It is located in the area of the **Orion arm of the Milky Way**, 28,000 light years from its center.
The primordial cloud from which the sun and the planets formed was several light years in diameter and had previously formed other first

THE SOLAR SYSTEM: THE SUN AND THE PLANETS

generation stars, which supplied heavier materials such as metals. More mass accumulated in the center, and it spun faster and faster. Near the sun, only metals could exist in solid form, since the gases evaporated, creating the **rocky planets:** Mercury, Venus, Earth and Mars, which could not be large, because these heavy elements were the least abundant. .

Far from the sun, where temperatures were lower, the light elements can be in a solid state, and since they were the most abundant, they formed the **giant gas planets**: Jupiter, Saturn, Uranus and Neptune.

When the thermal pressure was equal to the force of gravity, thermonuclear fusion of hydrogen began, which would last 10 billion years.

The Sun is the only object in the solar system that emits light due to the thermonuclear fusion of hydrogen which it transforms into helium. It has a diameter of 1,400,000 km, containing 99.8% of the mass of the solar system.

The solar wind is a current of plasma from the Sun, which crosses the solar system to its limits, in **the Oort cloud** one light year from the Sun.

THE SOLAR SYSTEM: THE SUN AND THE PLANETS

Planets and asteroids revolve around the Sun, following elliptical orbits, counterclockwise.

•**Interior or terrestrial planets**: Mercury, Venus, Earth and Mars.
•**Outer or giant planets:** Jupiter and Saturn (gas giants); Uranus and Neptune (frost giants)All giant planets have rings around them.

Dwarf planets have enough mass so that, due to the force of gravity, they have a spherical shape, but not to attract or expel all the objects around them.

Minor bodies of the solar system:
Asteroids, meteorites and comets.

Bodies that, without being a satellite, do not have enough mass to reach a spherical shape (about 800 km in diameter).
Except for the trans-Neptunian objects, the largest minor bodies in the

THE SOLAR SYSTEM: THE SUN AND THE PLANETS

solar system are Vesta and Pallas, with just over 500 km in diameter.
-**Asteroids** are smaller bodies located in an area between the orbits of Mars and Jupiter. Its size varies between 50 meters and 1000 km in diameter.
-**Meteoroids** are objects smaller than 50 meters in diameter, but larger than cosmic dust particles. They are usually fragments of comets or asteroids.
-**Satellites** are bodies that orbit around planets.

Beyond Neptune's orbit are: **the Kuiper belt** and **the Oort cloud,** where dwarf planets have been found.

Interplanetary space is not completely empty, there are particles of gas and dust from the evaporation of comets and the impacts of meteorites on the surface of the planets that, due to their weak force of gravity, cannot retain all the material. of the collision.
There are also energetic particles from the sun (**solar wind**). that reach the edge of the solar system (**heliopause**), located 100 times the distance from the sun to Earth.

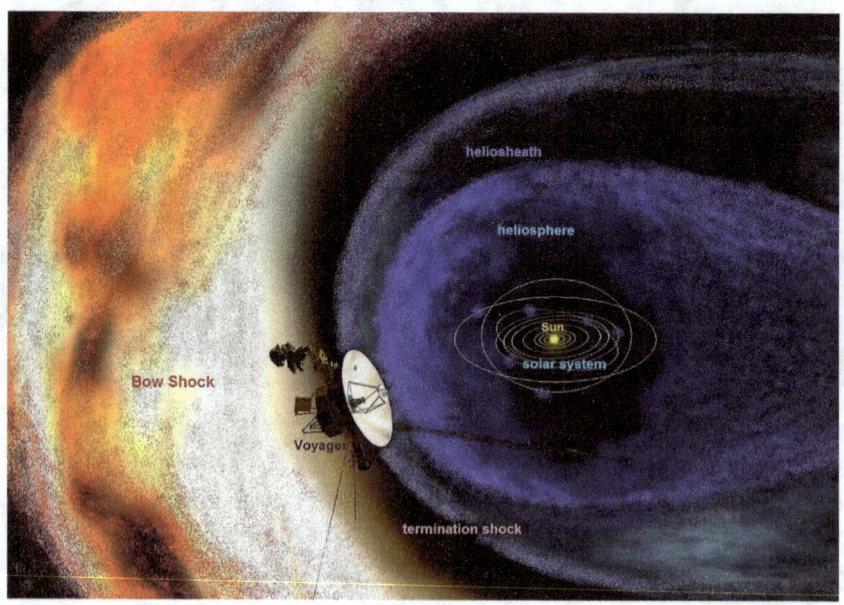

THE SOLAR SYSTEM: THE SUN AND THE PLANETS

SUN

The Sun is a sphere of plasma that generates a gigantic magnetic field. It is made up of 75% hydrogen.
The distance of the Sun from the Earth is 1 astronomical unit (150 million kilometers), 400 times the distance of the Moon, and its diameter is 109 times larger.
Every 11 years the Sun has a cycle of increased activity.
As in any other object in the universe, all the matter that composes it is attracted to the center by the force of gravity that its own mass generates.
The temperature at the center of the sun reaches 15 million degrees Celsius.

THE SOLAR SYSTEM: THE SUN AND THE PLANETS

Sunspots are areas where the temperature is lower than the rest. The temperature and gravitational pressure are so high that inside the stars, matter reaches a state that is neither gaseous, solid nor liquid, called plasma, the fourth state of matter.

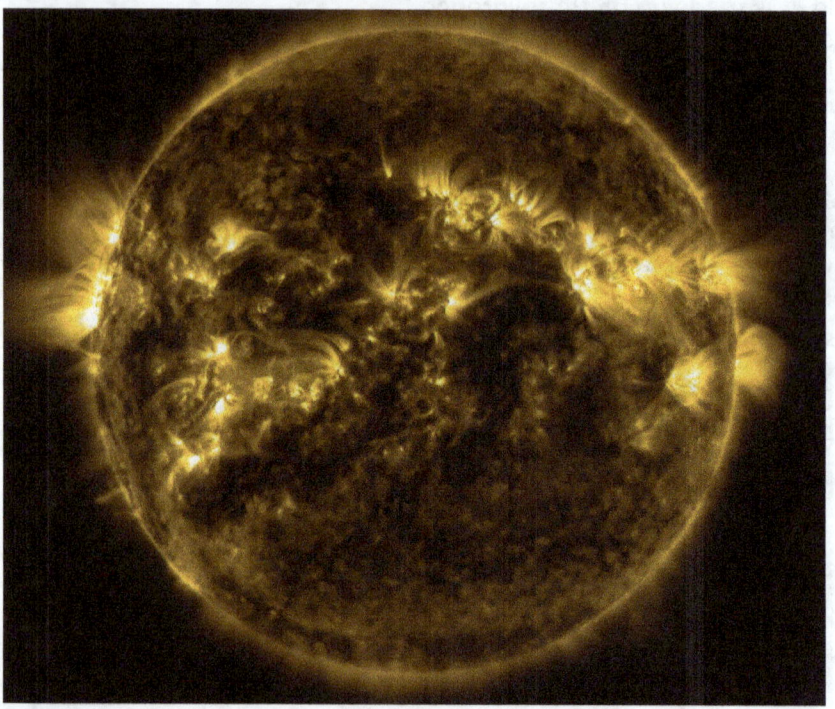

The sun transforms between 500 and 700 million tons of hydrogen into helium every second, and in this process, it expels more than 4 million tons of energy.
In fusion reactions there is a loss of mass, that is, the hydrogen consumed weighs more than the helium produced. That difference in mass is transformed into energy.
Solar radiation is estimated at 1000 Watts per m².

The energy generated in the core of the Sun takes a million years to reach the solar surface.
The intense force of gravity does not allow photons to escape, which

THE SOLAR SYSTEM: THE SUN AND THE PLANETS
creates **solar magnetism (solar wind).**

The solar wind drags gas and dust particles to the edge of the solar system. There, comets are formed with these materials, which return to the sun in an endless cycle.

When the sun consumes all the hydrogen, within 5 billion years, it will become a red giant star, increasing in size about 300 times and will begin to burn helium.
Then it will generate more energy than ever, melting the inner planets, expelling a large part of its mass in the form of a nebula, until the sun burns all the helium, and cools completely, becoming a white dwarf, one of the most dense of the Universe.

The Sun will not explode as a supernova because it does not have enough mass.

The combination of size and distance from the Sun and the Moon makes them appear the same apparent size.

White sunlight is made up of 7 colors: red, yellow, blue, green, indigo, orange and purple. When a ray of light passes through a raindrop at an angle of 40 degrees, it breaks down into all the colors that make up white, which are millions of shades, but the eye can only perceive a few.

COMETS

These are objects made up of rocks, ice and gases such as carbon dioxide and methane, which orbit around stars.

They also have organic compounds, the same as those that formed life on Earth, so some theories say that life originated from the collision of a comet.

There are more than 4,595 comets orbiting our sun. Although it is estimated that there could be more than a billion at the edge of the solar system, in the area called the Oort cloud.

-**The central part or core** can measure between 100 meters and 30 km in length.

-**Its hair or tail** can measure more than 150 million kilometers in length, the distance from the Earth to the Sun, and is made up of jets of gas and dust.

Comet Mc Naugth

As they approach the Sun, the cloud of gas and dust particles that surrounds them becomes charged with electricity thanks to the Sun's

THE SOLAR SYSTEM: THE SUN AND THE PLANETS

enormous magnetic field (solar wind), growing increasingly larger and forming the long hair of the comet.
The comet's hair of gas can be seen on the horizon just before dawn or after dusk, looking in the direction of the sun.

The first data on the observation of a comet date back to 230 BC.
In 44 BC, on the same day that the celebrations for the death of **Julius Caesar** began, a brilliant comment appeared over the skies of Rome and was visible in broad daylight, for 7 days in a row.
This fact was interpreted as a sign that Caesar's soul had ascended to heaven along with the other gods. His nephew Octavius Augustus spread this idea to support his candidacy for the government of Rome, and even built a temple where the comet was venerated.

In the year 66 B.C. The famous **Halley's Comet** was first seen, but it was not known when it would return. In 1705, astronomer **Edmund Halley** noted that the comet's orbit around the sun lasts 76 years and calculated that it would return in the year 1758, so the comet has been your name.

Comet Halley, Comet Hale-Bopp and Deep Impact

THE SOLAR SYSTEM: THE SUN AND THE PLANETS

In 1811, a comet was visible to the naked eye for 7 months, between March and September. It is believed to take almost 4,000 years to complete its orbit.

Most comets usually take between 20 and 200 years to return, although some take thousands of years, and others, like Comet Encke, return to the Sun every three years, but since it has lost almost all of its gas, it is no longer visible. naked eye.

Every time a comet passes near the Sun, it loses some of the gas in its tail. After about 2,000 orbits, it no longer has any gas left and becomes an asteroid.

In 1910, the tail of Halley's Comet measured 30 million kilometers, 1/5 of the distance between the Earth and the Sun.

When the Earth passes through the same area of space that a comet has already passed through, the small fragments that it broke off from its tail are attracted by the Earth's gravity and fall in the form of shooting stars.

METEORITES

What we call shooting stars are actually rocks of different sizes attracted by the force of gravity, called meteorites.

When falling at high speed, friction with the atmosphere causes the rock to reach such a high temperature that it shines in the sky for a few seconds, as if it were a star.

The Greek **Anaxagoras** already thought that meteorites were objects that arrived from the sun and were burning rocks.

Chladni, at the beginning of the 19th century, was the first scientist to accept that they had an extraterrestrial origin.

It is believed that more than 10,000 meteorites no larger than a soccer

THE SOLAR SYSTEM: THE SUN AND THE PLANETS

ball fall to the Earth's surface each year. Just over 5 are found. Before hitting the ground, most break down into particles smaller than grains of sand. However, larger meteorites can crash into the surface and form huge impact craters.

Meteorites brush against the atmosphere and reach temperatures above 2000 degrees Celsius.

Siliceous meteorites are close to 90% of the total, some come from of the origins of the solar system; others come from impacts of other meteorites against Mars and the Moon; metallic ones are less than 10%.

•**Metallic meteorites** (iron and nickel) melt more easily than rocky ones, as they are good conductors of heat, although they can reach the Earth's surface without breaking into millions of pieces.

•**Rocky meteorites** break into smaller and smaller fragments, until they completely disintegrate before reaching the ground, forming luminous trails, similar to fireworks.

Only those that are several kilometers in size can withstand the friction of the atmosphere and high temperatures.

-The largest meteorite found is **Hoba**, weighing 66,000 kg. It was discovered in 1920 in the Namibian desert and is believed to have fallen to Earth more than 80,000 years ago.

-**The ALH 84001 meteorite** comes from Mars and is 4.5 billion years old. 16 million years ago, the collision of a meteorite against the surface of Mars tore this rock from the planet, which overcame

Martian gravity, and traveling through the space, reached Earth 13,000 years ago.
-One of these meteorites fell 65 million years ago, in what is now the Yucatan Peninsule, Mexico, forming a huge crater and raising a cloud of dust and ash so large that it covered the Earth for years. This was the cause of the that the dinosaurs became extinct.
-Iron comes from stars much larger than the sun, which can melt helium and generate heavier metals, producing so much iron that gravity overcomes atomic force and the star collapses in the form of a supernova, expelling all those elements into the cosmos. items.

MERCURY

The Sumerians observed it 3000 years before Christ. The Babylonians called it the messenger of the gods, passing to Greece and Rome, who identify it with the god **Hermes/Mercury.**

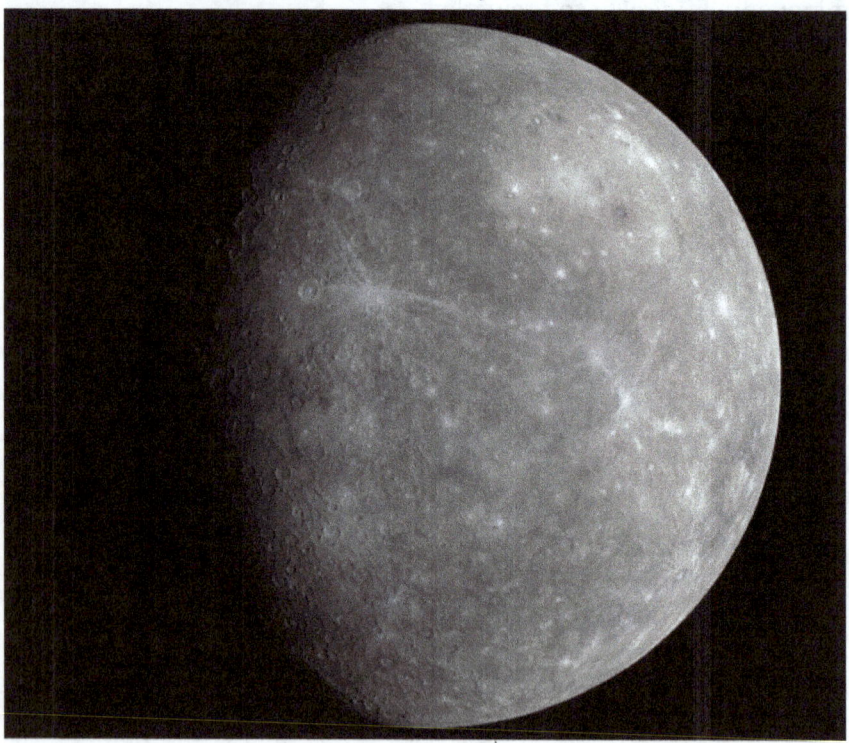

THE SOLAR SYSTEM: THE SUN AND THE PLANETS

Mercury can only be seen for a short period during sunrise and sunset. It is the smallest planet in the solar system, and the closest to the Sun. It is made up of rocks and has no atmosphere or satellites.

One day on Mercury is 58 Earth days. It takes 88 days to make a complete revolution around the sun.

Temperatures range between 623 K (350 °C) during the day and 103 K (−170 °C) at night. Ice has been found at the bottom of some craters. Just like on Earth, it has a magnetic field.

Curiously, it dawns and sets twice in that long day of 58 Earth days. The Sun rises and apparently stops in the sky, moving in the opposite direction.

VENUS

Called the morning star. It is named after the Roman goddess of love **(Venus/Aphrodite).**

It is the brightest object in the night sky, after the Moon.

THE SOLAR SYSTEM: THE SUN AND THE PLANETS

It can be seen three hours before sunrise or three hours after sunset. It is the second planet in the solar system closest to the Sun and the third in size, after Mars and Mercury. It has no satellites and its magnetic field is very weak.

It is a rocky planet and has one of the most spherical orbits. Temperatures reach 460 degrees Celsius, much higher than on Mercury, and, due to the thick cloud cover, there are hardly any thermal fluctuations.

The atmospheric pressure is 90 times higher than that of the earth (equivalent to the pressure 1000 meters deep in the ocean).

Its atmosphere is very dense and is composed of more than 90% carbon dioxide (CO_2), as well as nitrogen. Due to this high density, meteorites less than 3 km² do not reach its surface, and are completely disintegrated.

The clouds are composed of sulfur dioxide and sulfuric acid, with winds that reach 350 km/hour, in the highest layers of the atmosphere, and are more devastating than on Earth.

One day on Venus is equivalent to 243 days on Earth. In addition, the planet rotates in the opposite direction to the Earth, that is, from west to east, so there the Sun rises in the west and sets in the east.

The planet is covered by two extensive plateaus, separated by a plain.

THE SOLAR SYSTEM: THE SUN AND THE PLANETS

MOON

In Ancient Greece, **Anaxagoras** thought that the Sun and the Moon were two gigantic spherical objects and that the light of the Moon was the light of the sun reflected.
In 1609, Galileo observed the craters of the Moon.

It is thought that an object the size of Mars collided with the Earth, and from its remains, the Moon was formed.

THE SOLAR SYSTEM: THE SUN AND THE PLANETS

It is the fifth satellite in the solar system, its diameter: 3474.8 km, it is 1/5 of the diameter of the Earth.

The Moon rotates around the Earth at more than 3,600 km per hour, and since the orbit is not exactly circular, the closest distance to the Earth is 363,000 km, and the distance at which it is farthest from the Earth is 405,000 km.

The average distance between the Earth and the Moon is 384,000 km In the year 150 BC, **Hipparchus** already calculated the distance from the Earth to the Moon, with great precision.

The mass of the Earth is 80 times greater than that of the Moon, so gravity on the Moon is 6 times less than on Earth.

On Mars, gravity is half that on Earth, so an astronaut who weighs 100 kg on Earth will weigh 16.6 kg on the Moon and 50 kg on Mars.

On the Moon, an astronaut can jump up to 2.5 meters high.

THE SOLAR SYSTEM: THE SUN AND THE PLANETS

One day on the Moon is equivalent to almost 30 days on Earth. One night on the Moon is equivalent to almost 30 nights on Earth.
Since it takes the same time to rotate on its axis as it does to make a complete revolution around the Earth, counterclockwise, it always shows the same face or hemisphere, being able to see up to 60% of its surface.
The Sun always illuminates one half of the Moon.

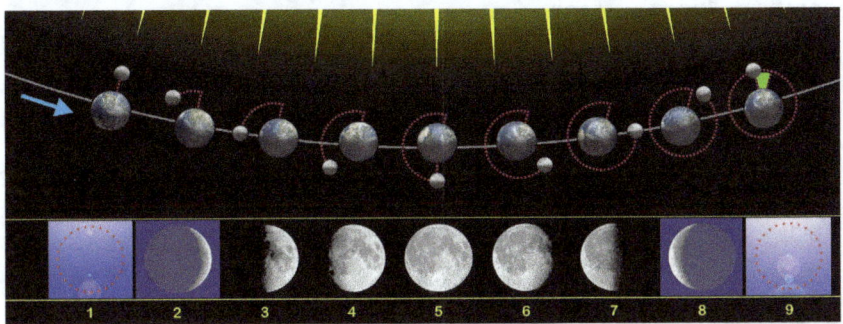

It is known that the Moon moves 4 centimeters away from the Earth per year, which gradually lengthens the duration of Earth's days, that is, it reduces the speed at which the Earth rotates.

-**Lunar eclipses** occur when the Earth comes between the Sun and the Moon, casting its own shadow that obscures the Moon.
The diameter of the Sun is 400 times larger than that of the Moon, but it is 400 times further away than the Moon, so the difference in sizes is compensated.

The moon has no magnetic field or atmosphere, which causes a large temperature fluctuation between day and night, reaching 120 degrees Celsius during the day and -230 degrees Celsius during the night.
The average temperature during the day is 100 degrees Celsius; and during the night it is -153 degrees Celsius.

Since it has no atmosphere, there is no wind, and its surface does not erode.
We can see the craters formed by asteroid impacts exactly the same as 3 billion years ago.

THE SOLAR SYSTEM: THE SUN AND THE PLANETS

It is thought that long before, it had geological activity, with numerous volcanic eruptions that created flatter surfaces, called seas.

More than 300 million tons of ice have been found in polar craters, since sunlight never reaches the interior and the temperature is always around -240 degrees Celsius. It can come from the impacts of comets or the solar wind. There are also extensive reservoirs of water beneath the lunar surface.
In 2013, a meteorite with a diameter of 1.4 meters and a weight of 400 kg collided in the area called the Sea of Clouds.

Trip to the Moon

Apollo 11 rocket launch

The Apollo missions took three days to reach the Moon.
When the first astronauts reached its surface, the temperature was 130 degrees Celsius.
They were protected by thick suits that weighed more than 130 kg and had 14 layers of insulation.

In 1969, the Apollo 11 mission landed the first man on the moon. The Apollo 11 mission computer, which controlled the command module, had just 4 kilobytes of RAM and 32 kilobytes of ROM, less memory than any old phone before smartphones.

The last manned mission was in 1972, with Apollo 17.
The Apollo 14 mission took 500 seeds of pine, spruce, sycamore and sequoia to the Moon, and exposed them to direct sunlight, to see what the effect of cosmic radiation was. They were then brought to Earth and

planted in various places, germinating more than 400 of these plants, called the trees of the moon.
In 2019, **China's Chang'e 4 mission** took cotton, potato, and rapeseed seeds to the moon and germinated them for a few days.
The Artemis mission will visit the moon between 2022 and 2034

Apollo 15 Lunar Rover

THE EARTH
The Earth rotates on its axis at a speed of 1,600 km/hour (rotational movement) and moves around the sun at 107,000 km/hour (translational movement). When it completes one revolution around the sun, it travels 930 million km.

THE SOLAR SYSTEM: THE SUN AND THE PLANETS

The Earth is not completely round, as it is 43 km wider at the equator than at the poles.
Light from the Sun takes 8 minutes and 17 seconds to reach Earth.

The African continent seen from space

The Earth has a magnetic field that protects it from cosmic rays or high-energy particles, which manage to pass through the heliosphere. The Earth's magnetic north pole is not located exactly in its geographical center, but about 1,600 kilometers away.

The **Moon's gravitational force** attracts everything on Earth. Very large objects such as masses of water are affected by this attraction, generating fluctuations in level, called tides.

THE SOLAR SYSTEM: THE SUN AND THE PLANETS

In a smaller body of water, such as a lake, there are also tides but they are so small that they cannot be seen with the naked eye, for example, in the Mediterranean they can be up to 30 centimeters between high tide and low tide.

The Moon's gravity also affects the Earth's spin.
4 billion years ago, the moon was 22,000 km away from Earth, and our planet was spinning very quickly.
1.4 billion years ago, a day lasted 18 hours.
Since then the Moon is gradually moving away from the Earth, causing it to rotate more slowly, so the days will last longer hours.
When in many millions of years, the Moon moves far enough away, the gravity it exerts will be so weak that the Earth's axis will vary in position, rotating around the equatorial zone, just as happens to Uranus.
Temperatures on the earth's surface range between 57 and -90 degrees Celsius, with winds that can exceed 200 km per hour.

The temperature difference between air masses generates **winds.** Hot air weighs less and rises; cold air weighs more and sinks.
Masses of very cold air form tiny ice crystals charged with electricity, and when they reach a certain level, an electric discharge or lightning occurs.
Most lightning strikes occur between clouds and do not reach the ground.

THE SOLAR SYSTEM: THE SUN AND THE PLANETS

-**Lightning** has an electrical charge of 15 million volts.
The current flow reaches 200,000 amperes.
The temperature reaches 30,000 degrees Celsius. The length of the beam is between 1.5 and 12 km, and they move in the air at a speed of more than 200,000 km per hour.
More than 2,000 storms form on Earth every day.

In Venezuela, at the mouth of the **Catatumbo River,** Lake Maracaibo area, storms develop every night between the months of April and November. The phenomenon has been taking place for 200 years, generating more than 10% of the earth's ozone.

Hurricanes form near the equator and move from east to west, in the same direction as the Earth's rotation, crossing the oceans.

-**Inside the earth**, the temperature is between 3,500 and 5,200 degrees Celsius and the pressure is 3.5 million times higher than that at sea level.
Below the Earth's crust is the mantle.
•Its upper part is made of solid materials, which stretch and contract without breaking.

THE SOLAR SYSTEM: THE SUN AND THE PLANETS

•Its lower part is formed by molten rocks and fluid materials, which generate magma currents due to differences in temperature and density:
•Hotter materials are less dense, weigh less and rise.
•Colder materials are denser, weigh more, and sink.

When these **magma** currents rise to the crust, they fracture it, forming plates through which heat, molten rocks and gases such as carbon dioxide escape.

Milky Way over Kilauea crater

The **magma** reaches a temperature of 1,200 degrees Celsius (2,100 degrees Fahrenheit) and can form a volcanic cone.

Most islands arose from the bottom of the oceans, from material expelled by underwater volcanoes.

Tectonic plates slide continuously or accumulate tension until they reach a level where slip occurs, causing an earthquake.
More than 500,000 earthquakes occur each year.

MARS

Rocky planet, it is the fourth most distant from the Sun and its size is half that of the Earth. Its name comes from the Greco-Roman god of war **Mars/Ares.**

It has a very tenuous atmosphere, with a pressure 100 times lower than Earth's, which is composed of 95% carbon dioxide, nitrogen and argon.

It has a core of iron, nickel and sulfur that is less dense than the earth's core, and whose gravity is 40%.

THE SOLAR SYSTEM: THE SUN AND THE PLANETS

The inclination of its axis of rotation is similar to the axis of the Earth, so on Mars, there are also seasons.
Mars takes 687 days to complete one revolution around the sun.
The day on Mars lasts 24 hours and 39 minutes.
One year on Mars is equivalent to 1 year and 10 months on Earth.

-The planet has the highest mountain in the solar system, Mount Olympus, 25 kilometers high, 600 km wide and a plateau that extends across 40% of the planet's surface.
-The great gorge called Valle Marineris has a length of 3000 km, a width of 600 km and a depth of 8 km.
-3/4 of Mars is covered by red rocks.

The average temperature is -55 ° degrees Celsius.
The minimum temperatures at the poles can drop to -130 degrees Celsius.
Daytime highs at the equator can exceed 20 degrees Celsius. while the nighttime minimums can drop to -80 degrees Celsius.

It had an ocean that covered 2/3 of the planet for 1.5 billion years.

THE SOLAR SYSTEM: THE SUN AND THE PLANETS

When the magnetic field of Mars disappeared, 4 billion years ago, the atmosphere escaped into outer space, lowering the pressure and temperature of the planet, and the water on the surface disappeared. With such low atmospheric pressure, water vapor goes from a gaseous state to a solid, in the form of ice, at a temperature of −80 degrees Celsius.

At the poles there is a permanent layer of CO_2 ice and water ice about 100 km long and 10 meters high.

Water ice clouds

The winds can blow at more than 150 km/hour and form extensive dune systems on its surface.

Sand storms can last for months and spread across the planet.

THE SOLAR SYSTEM: THE SUN AND THE PLANETS

There are white clouds composed of water vapor or carbon dioxide, and yellow clouds composed of microscopic sand particles, which give the sky a pinkish hue.
In winter, water vapor forms clouds of ice crystals and dry ice.

Mars has two small satellites, called Phobos and Deimos, whose orbits are very close to the planet. They come from the asteroid belt and have been captured by the planet's gravity.
Deimos, is the smallest and most distant. Phobos, is the largest and closest.
It takes less than 24 hours to orbit Mars, so it rises and sets in the sky twice a day.

Terrestrial planets sizes

THE SOLAR SYSTEM: THE SUN AND THE PLANETS

ASTEROID BELT

It is located between the orbits of Mars and Jupiter, at a distance between 2 and 4 astronomical units from the Sun.

It is made up of more than 500,000 asteroids with diameters greater than 1.5 km, and 1,000 asteroids with diameters greater than 15 km, in addition to extensive bands of cosmic dust of microscopic size, separated at great distances from each other.

They rotate around the sun in the same direction as the planets, taking between 3 years and 5 months, and 6 years to make a complete revolution.

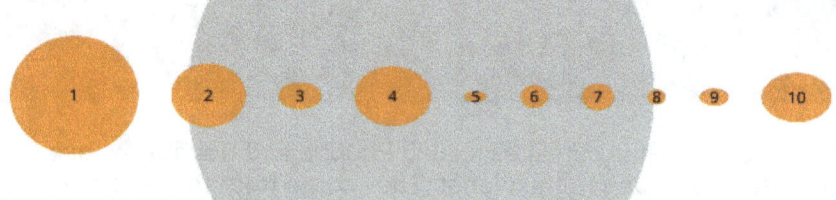

Moon 1 Ceres 2 Pallas 3 Juno 4 Vesta 5 Astraea 6 Hebe 7 Iris 8 Flora 9 Metis 10 Hygiea

Medium-sized asteroids are separated from each other by a distance of 5 million kilometers, so collisions occur at intervals of hundreds of thousands of years.

THE SOLAR SYSTEM: THE SUN AND THE PLANETS

Every 10 million years a collision occurs between asteroids whose radii are greater than 10 km. The collision causes the formation of smaller asteroids, if the speed is high; or the union of the two asteroids into one, if the speed is very low, which is less frequent.

The largest objects in the belt are **Ceres**, with 950 km, followed by **Pallas and Vesta**, with half the size.

Vesta, Ceres and Moon

The asteroid belt is thought to have formed at the same time as the planets in the solar system, 4.5 billion years ago.
During this early stage of formation of the solar system, these asteroids were unable to form a planet, as they were affected by the gravitational force of Jupiter.
•Some asteroids were accelerated so much in their trajectory that when they collided at high speed with others, they could not group together, and they divided into smaller and smaller fragments.
•Other asteroids extended their orbits around the sun so much that they collided with the sun, or were ejected into the Oort cloud, at the

edge of the solar system.
•Less than 1% of protoasteroids did not suffer significant collisions and retained their original shape.
The asteroids furthest from the sun conserve water and represent 75% of the total.
There are asteroids composed of iron and nickel, and even platinum.
1/3 of the asteroids travel around the sun, grouped with others, forming families, they come from the same asteroid that collided with another.

CERES

It is the largest object in the asteroid belt, considered a dwarf planet, one of the oldest planets or protoplanets, formed 4.5 billion years ago, along with **Vesta and Pallas**.

THE SOLAR SYSTEM: THE SUN AND THE PLANETS

It was discovered in 1801, and is named in honor of the Greco-Roman goddess of agriculture.

It has a diameter of 945 km and enough mass for it to be round in shape due to the force of gravity.

A day on Ceres lasts 9 hours, and it takes 4 years and 6 months to orbit the sun.

Its axis of rotation is inclined less than 4 degrees, so the polar areas are always subject to sunlight.

It is rocky and its surface is covered with ice. It is thought that there may be liquid water at great depths. It has geological activity, with some craters that expel dense brines.

The planet is full of craters that measure between 20 and 100 km wide and preserve a large amount of ice inside. The largest crater measures 280 km wide.

Occator crater

It has a very light atmosphere of water vapor generated by sublimation of surface ice.

Ceres has captured some asteroids for long periods of time, but has not cleared its orbit, which it shares with thousands of asteroids.

THE SOLAR SYSTEM: THE SUN AND THE PLANETS
- Ceres escape velocity: 0.51 km/s; 1836 km per hour.
- Moon escape velocity: 8640 km per hour.
- Earth escape velocity: 40,280 km per hour.

Escape velocity is what an object needs to escape the influence of another's gravitational field, for example, what a fragment of rock needs after the impact of an asteroid to escape the gravity of a planet and continue traveling through space. .

VESTA

Asteroid with a diameter of 530 kilometers, it has an iron and nickel core, and a basalt surface. It was discovered in 1807, named **in honor of the goddess of the home.** Its orbit is closer to the sun than Ceres. It rotates on its axis in just over 5 hours and takes 3 years and 6 months to make a complete revolution around the sun. Temperatures on its surface range between -20 and -130 degrees Celsius.

It had geological activity for a short period of geological time.
It has a crater 460 km in diameter at one of its poles, which measures between 4,000 and 12,000 meters high and 13 km deep.
It was caused by an impact from another object about 1 billion years ago.
Two other large impact craters measure more than 150 km wide and 7 km deep.

PALLAS

It was discovered after Ceres, in 1802, and is named in honor of **Pallas Athena, goddess of Wisdom.**
Pallas is 545 km in diameter, a similar size to Vesta, but with a lower density.
A day in Pallas lasts almost 8 hours. Its axis of rotation has more than 60° inclination, so sunlight hits it very unequally in winter and summer.

THE SOLAR SYSTEM: THE SUN AND THE PLANETS

JUPITER

It is the largest planet in the solar system, 318 times larger than Earth. It is located beyond Mars, the fifth due to its distance from the sun. It receives its name from the god **Jupiter/Zeus.**

It is part of the gaseous planets, made up of hydrogen and helium. Dense clouds cover the entire planet and the winds blow between 350 and 500 km/hour.

A day on Jupiter lasts 10 Earth hours.

Clouds are made of ammonia crystals and water vapor.

The high pressure of its atmosphere causes hydrogen to go into a liquid state, and then into a solid state. In the lower layers there is a large ice core with a size that is between 7 and 18 times that of the Earth.

It has the most intense magnetic field in the entire solar system.

THE SOLAR SYSTEM: THE SUN AND THE PLANETS

Diamonds can rain on Jupiter due to the very high pressure of its atmosphere. They are formed from carbon and descend from the upper layers to the lower layers.
-Jupiter has 67 **satellites**.In 1610, **Galileo** was able to observe its largest satellites: the volcanic Io, the icy Europa, the giant Ganymede, the largest satellite in the solar system, as well as Callisto, similar in appearance to our moon.

GANYMEDE
It is the largest satellite of Jupiter, with 5200 km in diameter, and one of the 4 that **Galileo** discovered in 1610. It was named in honor of **Jupiter's servant,** who was one of his lovers.

THE SOLAR SYSTEM: THE SUN AND THE PLANETS

It is twice as big as our Moon.
One day on Ganymede is equivalent to 7 days on Earth, which is also how long it takes to make a complete revolution around Jupiter, so it always presents the same face to the planet, just like our Moon.
It has a very tenuous atmosphere with small amounts of oxygen and hydrogen. It has a weak magnetic field.
It is composed of an iron core as well as silicon. Its surface is full of craters of different sizes and covered with a thick layer of ice.
It is divided into tectonic plates like on Earth, which millions of years ago formed the mountains. It no longer has geological activity. The extensive plain stands out.
Under its surface there is a gigantic ocean of liquid and salt water, larger in volume than on Earth.

CALLISTO

It is one of the four large satellites discovered by **Galileo**, the second largest of Jupiter, similar in size to Mercury.

Named after the **nymph, lover of Jupiter/Zeus**.
Its orbit is further away than that of the other three, and it always presents the same face to Jupiter, as happens with the Moon, to the Earth.

THE SOLAR SYSTEM: THE SUN AND THE PLANETS

One day on Callisto is equivalent to 17 days on Earth, and is also the time it takes to make a complete revolution around Jupiter, so it always presents the same face or hemisphere.
Rocky satellite with numerous craters without activity, a light carbon dioxide atmosphere and a strong magnetic field.
It has an ocean of frozen water more than 150 km deep and about 200 km thick.
It is known that the melting point of ice decreases with pressure, reaching -22 degrees Celsius when the pressure reaches 2,070 bar.
The flat surface is riddled with craters caused by meteorite impacts, which are of all sizes, being the one with the most craters in the entire solar system.

IO
It is the third satellite of Jupiter by size, and the closest of those discovered by **Galileo**.

THE SOLAR SYSTEM: THE SUN AND THE PLANETS

Rocky planet with mountains higher than those on Earth. Named after Io, she was a nymph, **lover of Jupiter/Zeus**, according to Greek mythology.

It is the planet in the solar system with the largest number of active volcanoes, more than 400. Clouds have been observed from eruptions that reach more than 500 km in height, which are attracted by Jupiter. On the surface there are lakes of liquid sulfur.

EUROPA

It is the smallest of the four satellites that **Galileo** discovered.
Its size is somewhat smaller than the Moon. Europa is the **mother of King Minos of Crete, lover of Jupiter/Zeus.**

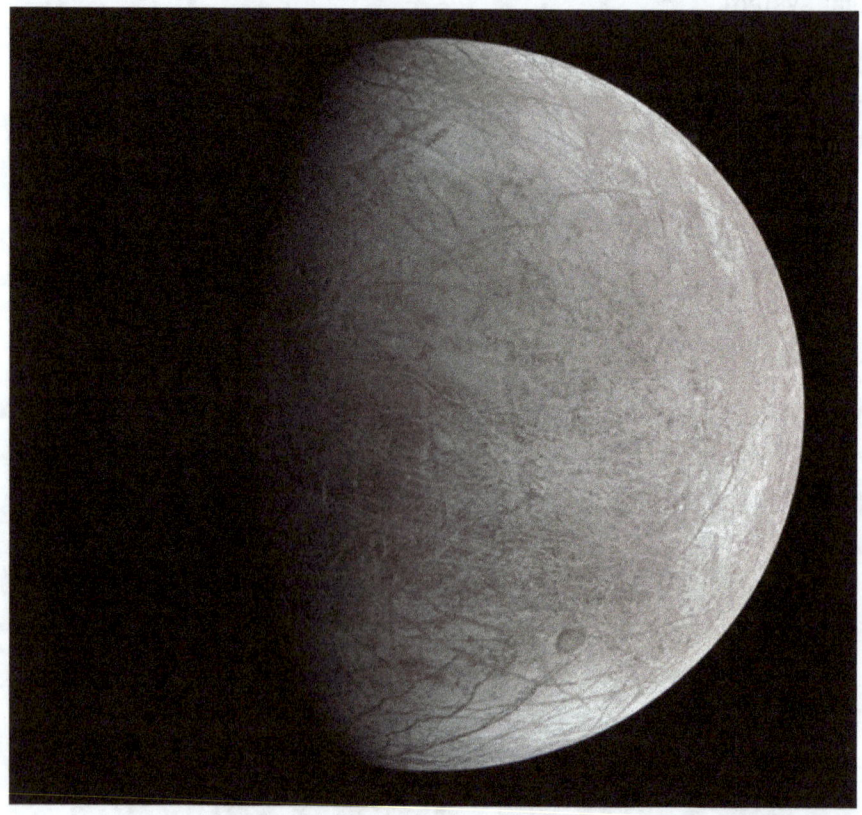

THE SOLAR SYSTEM: THE SUN AND THE PLANETS

It has an atmosphere rich in oxygen but very tenuous, although somewhat denser than that of Mars.
Temperatures range between -160 and -220 degrees Celsius.
Its interior is made of iron and nickel. At a depth of 25 km there is a thick layer of ice that surrounds the planet. At a depth of 150 km there is an ocean of salt water.

The Gallilean Satellites

SATURN

Gaseous planet whose name derives from the Greco-Roman god **Saturn/Cronus, son of Uranus and Gaea, and father of Jupiter/Zeus.**
It is 96 times larger than the Earth. Its atmosphere is composed of hydrogen and helium.
A day lasts just over 10 hours. The planet takes almost 30 years to make a complete revolution around the sun.
The high pressure and very high temperature, close to that found in the sun, means that these gases are in a liquid state.
Storms can last more than 7 months, and the lightning strikes have voltages of millions of degrees.
The magnetic field is much weaker than Jupiter's.

Saturn is surrounded by a vast belt. Although **Galileo** was the first to observe Saturn with a telescope, it was not until 1659 that **Christiaan Huygens** could clearly see its rings.
-The planet is surrounded by 1,000 rings made up of pieces of ice of different sizes that move at a speed of 48,000 km/hour.
Most are smaller than grains of sand and form a belt-shaped cloud of particles illuminated by sunlight. There are also some pieces the size of a truck or a house.
-There are 4 main ring stripes: A, B, C and D.
The rings measure between 100 meters and 400,000 kilometers wide, a distance greater than that between the Earth and the Moon.

THE SOLAR SYSTEM: THE SUN AND THE PLANETS

These rings are separated from each other by a spatial distance. They formed 100 million years ago, when dinosaurs inhabited the Earth. A huge comet collided with Saturn's atmosphere and disintegrated into millions of ice particles. Other scientists think they were formed by the collision of two of their icy moons.

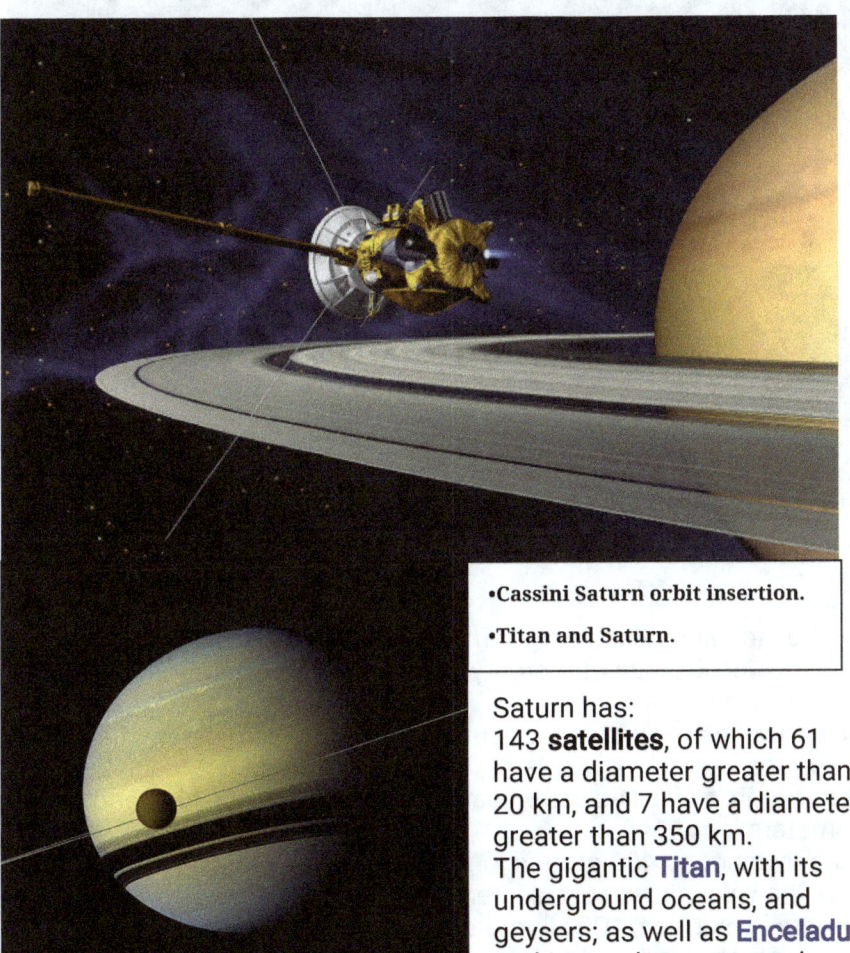

•Cassini Saturn orbit insertion.

•Titan and Saturn.

Saturn has:
143 **satellites**, of which 61 have a diameter greater than 20 km, and 7 have a diameter greater than 350 km.
The gigantic **Titan**, with its underground oceans, and geysers; as well as **Enceladus** and its methane atmosphere.
Huygens also discovered the satellite Titan.

THE SOLAR SYSTEM: THE SUN AND THE PLANETS

TITAN

It is the largest satellite of Saturn, with 5100 km in diameter, it is almost twice the size of Mercury. It is located 9.5 astronomical units from the sun.
It is a rocky planet with an ice surface and a weak magnetic field.
Its surface has extensive plains, mountains that do not reach 2000 meters high, as well as brown sand dunes 150 meters high and 1500 kilometers long.
It has rivers up to 400 meters long, and lakes of liquid methane at its poles. Its volcanic activity is intense.
Beneath its surface, 100 kilometers deep, there is an underground ocean of water and liquid ammonia.
The hydrocarbon reserves on this planet are thousands of times greater than those on Earth.

THE SOLAR SYSTEM: THE SUN AND THE PLANETS

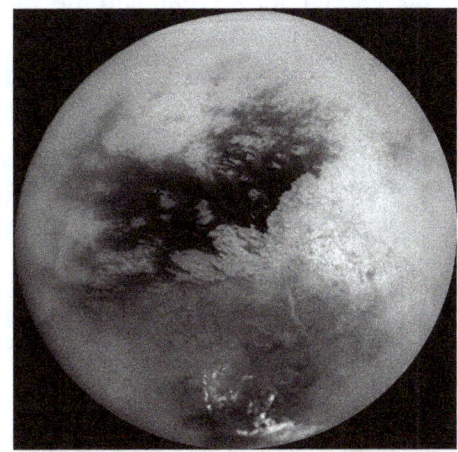

It has a dense atmosphere composed of 90% nitrogen and 5% methane, whose pressure is 1.5 times that of Earth.
The winds reach speeds of up to 180 km/hour.
The clouds reach up to 25 km high, although some can reach up to 100 km high.
Liquid methane rains on Titan, which on Earth is a gas, falling up to 50 liters per square meter per year. When it dries on the ground, it forms a layer of tar.

Rhea

THE SOLAR SYSTEM: THE SUN AND THE PLANETS

Most of the methane fallout evaporates again before reaching the ground.
On Titan, a day lasts 16 Earth days, the same time it takes to make a complete revolution around Saturn.
The sunlight that reaches Titan is 1000 times less than that which reaches Earth, being similar to dusk with a strong storm, so its surface temperature does not exceed -180 degrees Celsius.

RHEA

Saturn's satellite, the second largest after Titan, with more than 1500 km in diameter, half that of the Moon.
It was discovered in 1670 by astronomer Giovanni Cassini, and is named in honor of Rhea, **wife of Saturn/Cronus.**
It takes just 4 days to make a complete revolution around Saturn, although its orbit is very far from the planet.
It is composed of rock and ice. The surface is covered with craters.
It has a very light atmosphere of carbon dioxide and oxygen.
The temperature reaches -220 degrees Celsius.

IAPETUS

The third satellite of Saturn, due to its size, after Rhea and Titan. Named in honor of one of the Titans of mythology. It was discovered by **Giovanni Cassini** in 1671. It takes 79 days to make a complete revolution around Saturn (translational movement).

ENCELADUS

It is the sixth satellite by size, of Saturn with just over 500 km in diameter. It was discovered in 1789 by William Herschel.
It is a rocky planet whose surface is covered with ice. It has geological

activity. With hundreds of geysers, strips of land more than 100 km long that release jets of water vapor, salt crystals and ice. As the temperatures are - 198 degrees Celsius, part of the water they expel freezes quickly and falls to the ground in the form of snow; Another part is attracted by Saturn's gravity and adds material to its outer ring. Under its icy surface, 40 km deep, there is an ocean of salt water, which must have a high temperature, due to the geothermal activity of the satellite; which means very favorable conditions for life.

THE SOLAR SYSTEM: THE SUN AND THE PLANETS

It rotates rapidly around Saturn within the planet's outermost ring, in its narrowest area, taking 32 hours to make a complete revolution (translational movement).

It always presents its same face to Saturn, just like our Moon does to the Earth.

The south pole is surrounded by clouds of water vapor, with small amounts of nitrogen and carbon dioxide.

THE SOLAR SYSTEM: THE SUN AND THE PLANETS

PHOEBE

It is a Saturn-like satellite that does not have enough mass for gravity to give it a round shape, since its diameter is 220 km.

One day on Phoebe is equivalent to 9 hours. It takes 550 days to make a complete revolution around Saturn, which it does in the opposite direction to the rest. It is composed of ice and rock. Its surface is full of craters caused by asteroid impacts. The temperature is -163 degrees Celsius.
It is thought to have come from beyond Pluto, and that in its wandering through space it was trapped by Saturn's gravitational field.

URANUS

Farther than Saturn from the Sun is Uranus. It is the seventh planet in the solar system and the third in size, after Jupiter and Saturn. It is 63 times larger than the Earth.

Uranus is the **father of Saturn/Cronus and the grandfather of Jupiter/Zeus.**

It was discovered in 1781 by **William Herschel.**

Solar radiation is 400 times less than that which reaches the earth. The day lasts 17 Earth hours (rotation). It takes Uranus 84 years to orbit the Sun.

Its strange axis of rotation means that the planet's poles are located where the equator line is on Earth. This means that the poles have cycles of more than 40 years of light, and another 40 years of total darkness.

It has a magnetic field, as well as rings fainter than those of Saturn, and numerous satellites.

It has no solid surface. The atmosphere is mostly composed of hydrogen, in addition to helium and methane that mixes with the lower liquid layers, composed of water and ammonia, compressed by the very high pressure.

THE SOLAR SYSTEM: THE SUN AND THE PLANETS

Temperatures reach -200 degrees Celsius.
The winds on Uranus can reach up to 820 km/hour.
Uranus has a system of rings formed by microscopic pieces of ice, although some measure up to 1 meter in length, similar to that of Saturn.

Uranus Earth size comparison

-Uranus has 27 **satellites**, whose names come from characters in the works of **William Shakespeare**.
-It has five **major satellites**: Titania, Miranda, Oberon, Ariel and Umbriel. The smallest is Miranda with 470 kilometers, and the largest is Titania, with 1578 kilometer.
-Due to the great inclination of Uranus's axis of rotation, which causes one of its poles to always face the Sun, as its satellites rotate around Uranus's equator, the satellite poles also go through 42 years of darkness and 42 years of continuous light.
-All satellites are made of rock and ice, except Miranda, which is made of ice and carbon dioxide.

TITANIA

It is the largest of the satellites of Uranus. It was discovered by **William Herschel** in 1787. It is named after the **queen of the fairies** (A Midsummer Night's Dream, by William Shakespeare).
It has a weak carbon dioxide atmosphere, similar to that of Callisto, and much lighter than that of Pluto.
Its interior is rocky and the surface is covered with ice, under which it is believed that there is an ocean of liquid water that can reach 190 km deep.

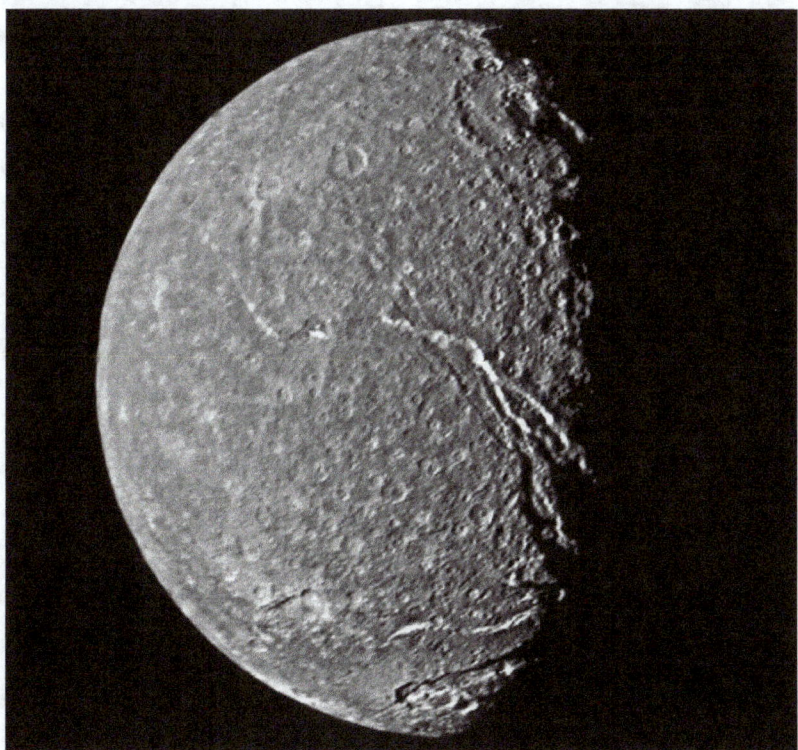

One day on Titania is equivalent to 8 days on Earth. The satellite always presents the same face to Uranus, just like our Moon does to Earth.
You can see numerous craters, gorges and plains.

MIRANDA

It is the smallest of the large satellites of Uranus, with 470 km in diameter. It was discovered in 1948, and is named in honor of the **daughter of the magician Prospero** (William Shakespeare's The Tempest).

Its interior is rocky with pockets of methane gas. Its surface is crossed by gorges and covered by water ice (you should know that other chemical elements also freeze, such as carbon dioxide...).

THE SOLAR SYSTEM: THE SUN AND THE PLANETS

OBERON

It is the second largest, after Titania, and the furthest from the main satellites of Uranus. Discovered in 1787, it is named in honor of Oberon, **the king of the Fairies** (A Midsummer Night's Dream, by William Shakespeare) . One day in Oberon is equivalent to almost 14 Earth days.
The satellite always presents the same face to Neptune, just like our Moon does to Earth, so it also takes 14 days to make a complete revolution around Uranus.
It is made of rocks and ice. It may have liquid water inside.
Its surface is completely covered by craters created by the impact of meteorites against its surface, some measuring more than 200 km. There are also deep gorges.
There are very dark areas, because the impacts of meteorites break the ice layer, exposing the rocky interior of Oberon.

NEPTUNE

It is the furthest planet from the sun.
It is named in honor of **Neptune/Poseidon, God of the sea.**
It is 17 times larger than the earth.
The disturbance in the orbits of Uranus and Saturn led mathematicians to assume that there must be another object beyond that was located in 1846 by Galle.
The atmosphere is made up of clouds of hydrogen, helium and methane.
The methane crystals transform into diamonds that fall as rain.
Below these clouds, and without a clearly defined separation, is an ocean of water and ammonia, charged with electricity, with temperatures of more than 4500 degrees Celsius. In the deepest part of the planet, there is a core of molten rocks.

THE SOLAR SYSTEM: THE SUN AND THE PLANETS

Neptune Earth size comparison

The temperature on the planet's surface is −218 degrees Celsius.
The wind speed reaches 2200 km/hour, the highest known.
Neptune has 17 **satellites**. The largest is Triton, where icy nitrogen geysers and the lowest temperatures in the solar system have been observed: −235 degrees Celsius.
Its ring system is similar to that of Jupiter.

TRITON

It is the largest satellite of Neptune. It was discovered by William Lassell in 1846, and named in honor of the **son of Neptune/Poseidon**, the god of the sea.
Its atmosphere is almost non-existent. On its surface temperatures of −235 degrees Celsius are reached, the lowest in the solar system. Triton's rotational motion is in the opposite direction to that of Neptune (retrograde orbit), so it is believed to come from the **Kuiper belt,** and was captured by Neptune's gravitational force.

THE SOLAR SYSTEM: THE SUN AND THE PLANETS

Neptune and Triton

The unusual inclination of the axis of rotation causes the poles to occupy the equatorial zone, as happens with Uranus. The seasons last 82 Earth years.
Triton orbits Neptune in a nearly circular orbit.
The interior is rocky and the surface of the poles is made up of frozen nitrogen and methane.
It has geological activity. There are volcanoes that expel liquid nitrogen and methane several km high.
Gravity brings Triton closer to Neptune and accelerates its rotation, until Triton gets so close that it collapses and forms a gigantic ring around Neptune.

THE SOLAR SYSTEM: THE SUN AND THE PLANETS

NEREID

Satellite discovered in 1949, and named in honor of the Nereids, nymphs that accompany Neptune, god of the sea.
It is 360 km in diameter, and its surface is covered in ice
A day in Nereida lasts 11 hours.
The orbit around Neptune is extremely elongated. Its closest point to the planet is 1.3 million km, and its furthest point from Neptune is almost 10 million km.

Gas planet size comparisons

THE SOLAR SYSTEM: THE SUN AND THE PLANETS

PLUTO
It was discovered by **Clyde Tombaugh** in 1930, it is named in honor of **Pluto/Hades, god of the underworld.**
Pluto is located in the Kuiper belt, a region that is at a distance between 30 and 50 astronomical units from the Sun.

It takes 248 years to make a complete revolution around the sun.
For 20 years the orbit of Pluto crosses the orbit of Neptune, but due to their inclination, there is no possibility of collision.
One day on Pluto is equivalent to 6 days on Earth. The inclination of its axis of rotation means that the planet's equator is located at its two poles, just as happens in Uranus.

THE SOLAR SYSTEM: THE SUN AND THE PLANETS

On Pluto the luminosity of the Sun is 1000 times lower than on Earth, being similar to a full moon night.
Its atmosphere of nitrogen, carbon dioxide and methane is very tenuous. On its surface there is frozen methane and hydrogen.

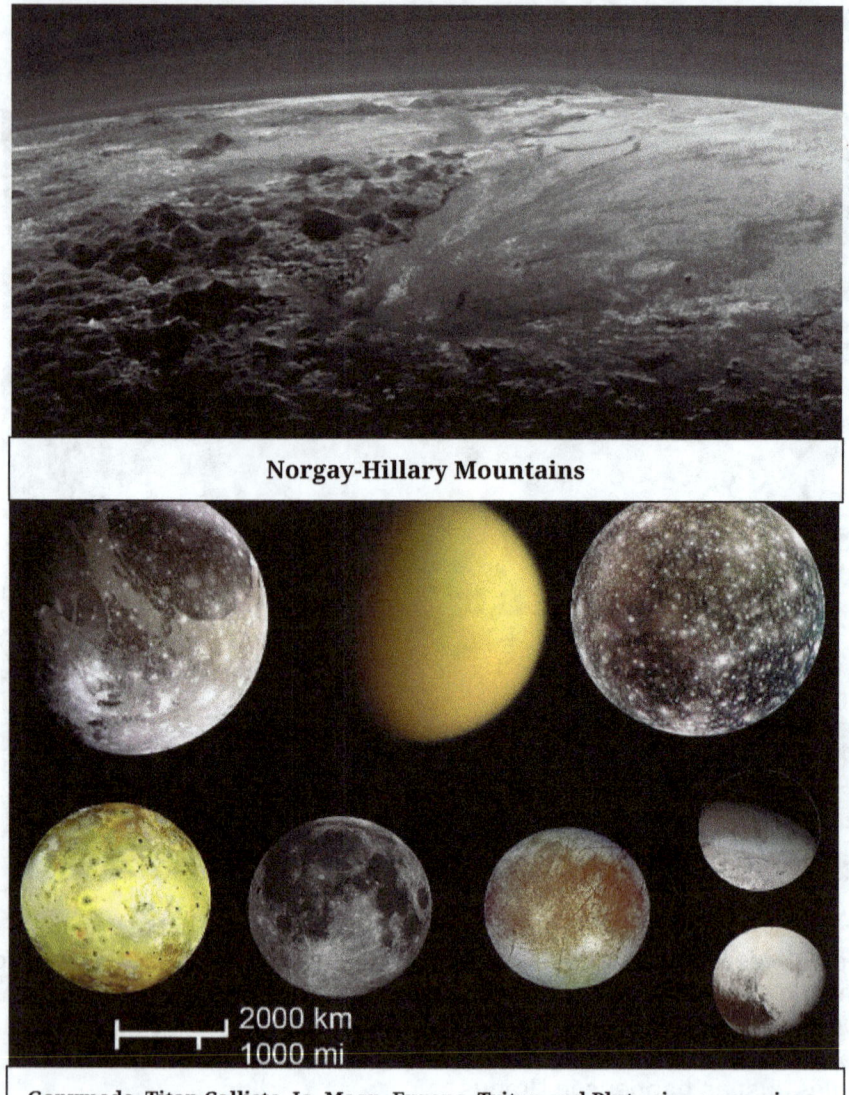

Norgay-Hillary Mountains

Ganymede, Titan Callisto, Io, Moon, Europa, Triton and Pluto size comparions

It has 5 **satellites**: Charon, discovered in 1978, which is similar in size to Pluto although much less mass; Nyx, Hydra, Cerberus and Styx.

CHARON
It is the largest satellite of Pluto and was discovered by **James W. Christy** in 1978. It is named in honor of Charon, a **boatman who was responsible for taking the souls of the dead to the underworld.**
It is 1,200 kilometers in diameter and is 19,000 kilometers from Pluto, 20 times closer than the Moon is to Earth.
Charon always presents the same face to Pluto, just as the Moon does to the Earth.
Its interior is rock and ice, and its surface is covered in water ice and lacks an atmosphere. The temperature ranges to -258 degrees Celsius.
Charon does not revolve around Pluto like a satellite, but rather Pluto and Charon revolve around a common gravitational point, (double planetary system).

DWARF PLANETS BEYOND NEPTUNE

-In 2002 and 2003, **Quaoar** and **Sedna** were discovered, whose diameter is half the diameter of Pluto.
-In 2005, **Eris** was discovered, with a greater mass than Pluto.
-**Hydra** is 55 km and **Nix** is 42 km. Both have no axis of rotation, so they rotate chaotically.
-**Cerberus** and **Styx** were discovered in 2011 and are about 11 km in diameter.

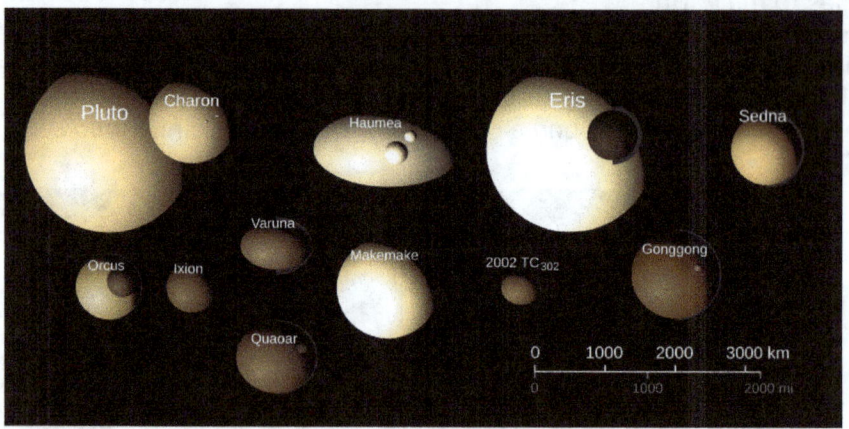

ERIS

It is the trans-Neptunian dwarf planet with the most mass, and the second largest after Pluto, with 2,300 km in diameter.

It was discovered in 2005 by the Mount Palomar observatory in the United States of America.

It was named in honor of the **goddess of discord, who caused the Trojan War.**

Its interior is rocky and the surface is composed of frozen methane.
Its orbit around the sun is 3 times farther than Pluto's orbit.
It takes 557 years to orbit the sun, which is between 35 and 95 astronomical units away.
Pluto orbits the sun between 29 and 49 astronomical units away.

THE SOLAR SYSTEM: THE SUN AND THE PLANETS

Neptune orbits the sun at 30 astronomical units.
It has a **satellite** called **Dysnomia**, goddess of unjust actions.

SEDNA

Located in **the Oort cloud**, between 76 and 960 astronomical units from the sun, about 32 times farther than Neptune.
It was discovered in 2003 by the Mount Palomar observatory in the United States. It is named in honor of the **Eskimo goddess of the sea.**
Its diameter is 1600 km. A day in Sedna lasts 10 hours.
It takes 11,400 years to orbit the sun. A spacecraft would take almost 25 years to reach this object.
Its surface is composed of carbon ice, methane and frozen nitrogen. The temperatures are lower than -230 degrees Celsius so it is believed that methane does not evaporate and then fall as snow as happens on Triton and Pluto.

MAKEMAKE

Dwarf planet located in the Kuiper belt. A spacecraft would take 16 years to arrive. It was discovered in 2005. It is named in honor of a **deity from Easter Island.**
It takes 308 years to orbit the sun. Its size is 1450 km in diameter, 60% of Pluto.

THE SOLAR SYSTEM: THE SUN AND THE PLANETS

Its surface is covered with ice, nitrogen and frozen methane.
It is believed to have a slight atmosphere of nitrogen and methane.
It has a **satellite** 21,000 km away, which is 175 km in diameter, and takes 12 days to orbit Makemake.

HAUMEA

Ellipse-shaped dwarf planet, located in the **Kuiper belt**.
It was discovered in 2003, and was named in honor of the Hawaiian goddess of fertility. It is 1/3 the size of Pluto, about 1400 km in diameter, and has rings around it.
It is located 35 astronomical units from the sun. It rotates around itself in 4 hours, and takes 283 years to make a complete revolution around the sun.
It is a rocky planet whose surface is covered in ice.
It is believed to have no atmosphere.

It has two **satellites**, the largest, called **Hi'iaka**, in honor of the Hawaiian goddess of medicine, is the outermost, is located 50,000 km away, has a diameter of 300 km and takes 49 days to orbit the planet. . The youngest is called **Namaka**, in honor of the Hawaiian goddess of the sea.

QUAOAR

Dwarf planet candidate, located in the distant Kuiper belt, at the edge of the solar system. It was discovered in 2002 by the Palomar Mountain observatory.

THE SOLAR SYSTEM: THE SUN AND THE PLANETS

Named in honor of a god of the first inhabitants of North America. It has a diameter of 1100 km, half the size of Pluto, and has a system of two rings, made up of ice fragments up to 300 km wide.
Its surface is covered with ice.
A **satellite** called **Weywot** rotates around it.

GONGGONG

In 2007 by the Mount Palomar Observatory. Named after the **Chinese god of the sea.**
It has a diameter of 1200 km, and a **satellite** called **Xiangliu**.
It takes 553 years to orbit the sun.
Its surface is believed to be covered with water ice and, perhaps, frozen methane.

THE SOLAR SYSTEM: THE SUN AND THE PLANETS

THE SOLAR SYSTEM: THE SUN AND THE PLANETS

Copyright2024. The Solar System: The Sun and the Planets.Published by Baltasar Rodríguez Otero at Kindle.

Acknowledgements
-https://upload.wikimedia.org/wikipedia/commons/c/c5/Released_to_Public_Voyager_Mo
ntage_by_NASA_(NASA)_(291707648).jpg
Released to Public: Voyager Montage by NASA (NASA) Author pingnews.com
https://upload.wikimedia.org/wikipedia/commons/3/31/Sizes_of_Solar_System_objects_to
_scale.png23 January 2024 Source Own work Author RedKire25
https://upload.wikimedia.org/wikipedia/commons/thumb/5/51 High_School_Earth_
Science_Cover.jpg/http://cafreetextbooks.ck12.org/science/CK12_Earth_Science.pdf
If the above link no longer works, visit http://www.ck12.org and lookfor the CK-12 Earth
Science book.Author CK-12 Foundation
https://upload.wikimedia.org/wikipedia/commons/thumb/2/20/Nh-pluto-charon-v2-10-1-15
_1600.jpg/NASA Solar System Exploration Author NASA's New Horizons
spacecraft
https://commons.m.wikimedia.org/wiki/File:Solar_sys.jpghttps://photojournal.jpl.nasa.gov/
catalog/PIA11800Author NASA/JPL
https://upload.wikimedia.org/wikipedia/commons/7/7e/Solar_system_Painting.jpg Harman
Smith and Laura Generosa (nee Berwin), graphic artists and contractors to NASA's Jet
Propulsion Laboratory.
https://upload.wikimedia.org/wikipedia/commons/thumb/d/de/The_Solar_System_(373075
79045).jpg/The Solar System Author Kevin Gill from Los Angeles, CA, United States
https://upload.wikimedia.org/wikipedia/commons/thumb/f/f0/2006-16-d-print2.jpg/1078px
-2006-16-d-print2.jpg Source Page:http://hubblesite.org/newscenter/newsdeskarchive/
releases/2006/16/image/dAuthor A. Feild(SpaceTelescope Science Institute)From
http://hubblesite.org/copyright/ copyright@stsci.edu.
https://upload.wikimedia.org/wikipedia/commons/thumb/a/af/NASA_Heliosphere_Mod.jpg
/NASA/JPL-Caltech.Author JudithNabb
https://upload.wikimedia.org/wikipedia/commons/thumb/b/b7/Asteroid_Bennu's_Journey%
2C_the_formation_of_our_Solar_system_and_the_early_Earth_(NASA_video).webm/.jpg
NASA | Asteroid Bennu's Journey −View/savearchivedversions on archive.org and archive
today Author NASA Goddard
https://upload.wikimedia.org/wikipedia/commons/thumb/0/0b/BENNU'S_JOURNEY_-_Early
_Earth.jpg/Flickr Author NASA's Goddard Space Flight Center
https://upload.wikimedia.org/wikipedia/commons/8/81/Solar_System_Diagram_-_Feb._201
9_(46327506074).jpgStephenposted to Flickr by splinx1 at https://flickr.comphotos/
42837737@N05/46327506074
https://upload.wikimedia.org/wikipedia/commons/thumb/6/68/Artist's_conception_of_Sed
na.jpg NASA/JPL-Caltech/R. Hurt(SSC-Caltech)
https://upload.wikimedia.org/wikipedia/commons/thumb/3/38/Haumea_with_rings_(37641
832331).jpg/ Kevin Gill from Los Angeles,CA,UnitedStateshttps://flickr.com/photos/
53460575@N03/37641832331
https://upload.wikimedia.org/wikipedia/commons/thumb/b/bc/Artist's_concept_of_the_Sol
ar_System_as_viewed_from_Sedna.jpg/http://hubblesite.org/newscenter/archive/releases/
2004/14/image/f/formatlarge_web/Author NASA,ESA and Adolf Schaller

THE SOLAR SYSTEM: THE SUN AND THE PLANETS

https://upload.wikimedia.org/wikipedia/commons/thumb/2/21/10_Largest_Trans-Neptunian_objects_(TNOS).png/Lexicon(Commons 3.0),Exoplanet Expert (Commons 4.0),SpaceDude777

-https://upload.wikimedia.org/wikipedia/commons/thumb/c/c7/Saturn_during_Equinox.jpg/
http://www.ciclops.org/view/5155/Saturn-Four-Years-On
http://www.nasa.gov/images/content/365640main_PIA11141_full.jpg
http://photojournal.jpl.nasa.gov/catalog/PIA11141 Autor NASA / JPL / Space Science Institute

-https://upload.wikimedia.org/wikipedia/commons/thumb/9/97The_Earth_seen_from_Apollo_17.jpg/NASA/Apollo 17 crew; taken by either Harrison Schmitt or RonEvans

-https://upload.wikimedia.org/wikipedia/commons/thumb/0/01/Phase-180.jpg/Jay Tanner

-https://upload.wikimedia.org/wikipedia/commons/thumb/d/df/Full_moon_partially_obscured_by_atmosphere.jpg

http://spaceflight.nasa.gov/gallery/images/shuttle/sts-103/html/s103e5037.html Autor NASA

-https://upload.wikimedia.org/wikipedia/commons/thumb/4/44 Kilauea_Volcanic_Eruption_Big_Island_Hawaii_2018_(31212271237).jpg/Author Anthony Quintano from Mount Laurel, United States

-https://upload.wikimedia.org/wikipedia/commons/thumb/8/89/Comet_C-1995_O1_Hale-Bopp%2C_on_March_14%2C_1997_(cropped).jpg/Author ignoto - Credit: ESO/E. Slawik

-https://upload.wikimedia.org/wikipedia/commons/thumb/8/86/Montagem_Sistema_Solar.jpg/NASA

-https://upload.wikimedia.org/wikipedia/commons/thumb/3/3b/Portrait_of_Sir_Isaac_Newton%2C_1689.jpg/https://exhibitions.lib.cam.ac.uk/linesofthought/artifacts/newton-by-knellr

-https://upload.wikimedia.org/wikipedia/commons/thumb/d/d8/NASA_Mars_Rover.jpg/1280px-NASA_Mars_Rover.jpgNASA/JPL/Cornell University, Maas Digital LLC

https://upload.wikimedia.org/wikipedia/commons/thumb/6/68/Schiaparelli_Hemisphere_Enhanced.jpg

https://astrogeology.usgs.gov/search/details/Mars/Viking/schiaparelli_enhanced/tif Autor USGS

https://upload.wikimedia.org/wikipedia/commons/thumb/f/f6/May_28%2C_2013_Bennington%2C_Kansas_tornado.jpeg/Dustin Goble (Submitted to National Weather Service)

https://upload.wikimedia.org/wikipedia/commons/thumb/1/12/Oidipous_sphinx_MGEt_16541_reconstitution.svg/Juan José Moral.

https://upload.wikimedia.org/wikipedia/commons/thumb/b/b4/The_Sun_by_the_Atmospheric_Imaging_Assembly_of_NASA's_Solar_Dynamics_Observatory_-_20100819.jpg/NASA/SDO (AIA)

https://upload.wikimedia.org/wikipedia/commons/thumb/0/02/SolarSystem_OrdersOfMagnitude_Sun-Jupiter-Earth-Moon.jpg/Tdadamemd

https://upload.wikimedia.org/wikipedia/commons/thumb/f/f3/Orion_Nebula_-_Hubble_2006_mosaic_18000.jpg/NASA, ESA, M. Robberto (Space Telescope Science Institute/ESA) and the Hubble Space Telescope Orion Treasury Project Team

https://upload.wikimedia.org/wikipedia/commons/thumb/6/63/Messier_81_HST.jpg/NASA, ESA and the Hubble Heritage Team (STScI/AURA)

https://upload.wikimedia.org/wikipedia/commons/a/ae/EastHanSeismograph.JPGen:user:Kowloonese

THE SOLAR SYSTEM: THE SUN AND THE PLANETS
https://es.m.wikipedia.org/wiki/Archivo:TakakkawFalls2.jpg Michael Rogers (Mjrogers50 de Wikipedia en inglés)
https://upload.wikimedia.org/wikipedia/commons/thumb/8/85/Venus_globe.jpg/photojournal.jpl.nasa.gov/catalog/PIA00104Autor NASA/JPL
https://upload.wikimedia.org/wikipedia/commons/thumb/7/7c/Terrestrial_planet_sizes2.jpg/NASA/JHUAPLVenus image:NASA/Johns Hopkins University Applied Physics Laboratory/Carnegie Institution of Washington Earth image: NASA/Apollo 17 crew, retouch by User:Aaron1a12
https://upload.wikimedia.org/wikipedia/commons/thumb/7/71/PIA22946-Jupiter-RedSpot-JunoSpacecraft-20190212.jpg/NASA/JPL-Caltech/SwRI/MSSS/Kevin M. Gill
https://upload.wikimedia.org/wikipedia/commons/thumb/9/95/Uranus%2C_Earth_size_comparison_2.jpg/NASA (image modified by Jcpag2012)
https://upload.wikimedia.org/wikipedia/commons/thumb/2/2f/Neptune%2C_Earth_size_comparison_true_color.jpg/CactiStaccingCrane
https://upload.wikimedia.org/wikipedia/commons/thumb/1/1c/Europa_in_natural_color.png/Europa - PJ45-2.png from https://www.missionjuno.swri.edu/junocam/processing?id=13844 Autor NASA/JPL-Caltech/SwRI/MSSS/Kevin M. Gill
https://upload.wikimedia.org/wikipedia/commons/thumb/2/21/Ganymede_-_Perijove_34_Composite.png/2048px-Ganymede_-_Perijove_34_Composite.png Kevin M. Gill
https://flickr.com/photos/53460575@N03/51238659798 Ganymede -Perijove 34 CompositeAutor NASA/JPL-Caltech/SwRI/MSSS/Kevin M.Gill
https://upload.wikimedia.org/wikipedia/commons/thumb/0/0e/Moon_and_Asteroids_1_to_10.svg/Vystrix Nexoth
https://upload.wikimedia.org/wikipedia/commons/thumb/b/ba/Dawn_Flight_Configuration_2.jpg/jpghttp://dawn.jpl.nasa.gov/multimedia/spacecraft.asp GDKDawn spacecraft Source:http://dawn.jpl.nasa.gov/multimedia/spacecraft.asp PD-NASA
https://upload.wikimedia.org/wikipedia/commons/thumb/7/7b/Io_highest_resolution_true_color.jpg/NASA /JPL /University of Arizona
https://upload.wikimedia.org/wikipedia/commons/thumb/0/06/Titan_in_front_of_the_ring_and_Saturn.jpg/http://photojournal.jpl.nasa.gov/catalog/PIA14922 Author Produced By Cassini Credit:NASA/JPL-Caltech/Space Science Institute
https://upload.wikimedia.org/wikipedia/commons/thumb/2/25/Titan_globe.jpg/NASA/JPL/ Space Science Institute Permissionjpl.nasa.gov
https://upload.wikimedia.org/wikipedia/commons/thumb/b/b2/Cassini_Saturn_Orbit_Insertion.jpg/Autor NASA/JPL
https://upload.wikimedia.org/wikipedia/commons/4/46/Gas_planet_size_comparisons.jpg http://solarsystem.nasa.gov/multimedia/display.cfm?IM_ID=180Author Solar System Exploration, NASA
https://upload.wikimedia.org/wikipedia/commons/7/7d/PIA01482_Saturn_Montage.jpg JPL image PIA01482 Author NASA
https://upload.wikimedia.org/wikipedia/commons/thumb/d/d4/Justus_Sustermans_-_Portrait_of_Galileo_Galilei%2C_1636.jpg/identificador Art UK de unaobra de arte: galileo-galilei-15641642-175709
fotógrafo https://www.rmg.co.uk/collections/objects/rmgc-Dmitry Rozhkov object-14174
https://upload.wikimedia.org/wikipedia/commons/thumb/3/30/Mercury_in_color_-_Prockter07_centered.jpg/NASA/JPLAutor NASA /Johns Hopkins University Applied Physics Laboratory /Carnegie Institution of Washington.Prockter07.jpg by Papa Lima Whiskey .

THE SOLAR SYSTEM: THE SUN AND THE PLANETS

https://upload.wikimedia.org/wikipedia/commons/5/58/Ceres_-_RC3_-_Haulani_Crater_(22381131691).jpgCeres -RC3 -Haulani Crater Autor Justin Cowart

https://upload.wikimedia.org/wikipedia/commons/thumb/4/41/Sol454_Marte_spirit.jpg/http://marsrovers.jpl.nasa.gov/gallery/press/spirit/20050420a.html Autor NASA/JPL

https://upload.wikimedia.org/wikipedia/commons/thumb/f/f5/007_Jack's_4_O'clock_EVA-1_LM_Pan_Hi_Res.jpg/NASA/Gene Cernan/Jack Schmitt

https://upload.wikimedia.org/wikipedia/commons/thumb/8/8e/Duke_on_the_Descartes_-_GPN-2000-001123.jpg/Author NASA John Young

https://upload.wikimedia.org/wikipedia/commons/thumb/e/e4/Water_ice_clouds_hanging_above_Tharsis_PIA02653_black_background.jpg/http://www.jpl.nasa.gov/spaceimages/details.php?id=PIA02653 Author NASA/JPL/MSSS

https://upload.wikimedia.org/wikipedia/commons/thumb/c/cb/7505_mars-curiosity-rover-gale-crater-beauty-shot-pia19839-full2.jpg/https://mars.nasa.gov/resources/7505/Author Jim Secosky picked out a NASA JPL-Caltech

https://commons.m.wikimedia.org/wiki/File:Lspn_comet_halley.jpg NASA/W.Liller

https://upload.wikimedia.org/wikipedia/commons/thumb/0/0c/360°_View-_Very_Well-Preserved_9-Kilometer_Diameter_Impact_Crater_(33432247000).jpg/https://flickr.com/photos/53460575@N03/33432247000Author Kevin M. Gill Flickr set Hourly Cosmos Flickr

https://upload.wikimedia.org/wikipedia/commons/thumb/f/f9/Ceres_and_Vesta%2C_Moon_size_comparison.jpg/Gregory H. Revera Ceres image: Justin Cowart Vesta image: NASA/JPL-Caltech

https://upload.wikimedia.org/wikipedia/commons/thumb/f/f9/Sar2667_as_it_entered_Earth's_atmosphere_over_the_north_of_France.jpg/Wokege

https://upload.wikimedia.org/wikipedia/commons/thumb/5/5a/Uranus_moons.jpg/Vzb83

https://upload.wikimedia.org/wikipedia/commons/thumb/e/e1/HAVO_20220213_Milky_Way_over_Kilauea_crater_J.Wei_(51888623142).jpg/Hawaii Volcanoes National Park

https://upload.wikimedia.org/wikipedia/commons/thumb/3/3b/Catatumbo_Lightning_-_Rayo_del_Catatumbo.jpg/Fernando Flores from Caracas,Venezuela https://flickr.com/photos/44948457 @N07/23691566642

https://es.m.wikipedia.org/wiki/Archivo:Huracan_patricia_23-10.jpghttps://twitter.com/StationCDRKelly/status/657618739492474880Autor Scott Kelly

https://es.m.wikipedia.org/wiki/Archivo:PIA17202_-_Approaching_Enceladus.jpg National Aeronautics and Space Administration (NASA) Jet Propulsion Laboratory (JPL)

https://commons.m.wikimedia.org/wiki/File:Callisto_-_May_26_2001_(37113416323).jpg Kevin Gill from Los Angeles, CA, United States Flickr by Kevin M. Gill at https://flickr.com/photos/53460575@N03/37113416323

https://commons.m.wikimedia.org/wiki/File:The_Galilean_Satellites_-_PIA01299.tiffJPLAuthor NASA

https://commons.m.wikimedia.org/wiki/File:PIA00340_Montage_of_Neptune_and_Triton.jpg http://photojournal.jpl.nasa.gov/ catalog/PIA00340 Author NASA,JPL

https://upload.wikimedia.org/wikipedia/commons/thumb/e/ef/Pluto_in_True_Color_-_High-Res.jpg/1024px-Pluto_in_True_Color_-_High-Res.jpgNASA/Johns Hopkins University Applied Physics Laboratory/Southwest Research Institute/Alex Parker

https://upload.wikimedia.org/wikipedia/commons/thumb/c/c9/Iapetus_as_seen_by_the_Cassini_probe_-_20071008.jpg/The Other Side of Iapetus Autor NASA / JPL / Space Science Institute

THE SOLAR SYSTEM: THE SUN AND THE PLANETS

https://upload.wikimedia.org/wikipedia/commons/thumb/2/23/Pluto_compared2.jpg/Composition of NASA images by Eurocommuter.
https://upload.wikimedia.org/wikipedia/commons/thumb/a/a3/PIA19947-NH-Pluto-Norgay-Hillary-Mountains-20150714.jpg/NASA/Johns Hopkins University Applied Physics Laboratory
https://upload.wikimedia.org/wikipedia/commons/thumb/2/2e/Charon_in_True_Color_-_High-Res.jpg/NASA/Johns Hopkins University Applied Physics Laboratory/Southwest Research Institute/Alex Parker
https://upload.wikimedia.org/wikipedia/commons/thumb/a/ab/PIA07763_Rhea_full_globe5.jpg/http://photojournal.jpl.nasa.gov/catalog/PIA07763Autor NASA /JPL/Space Science Institute
https://upload.wikimedia.org/wikipedia/commons/thumb/2/21/Ganymede_-_Perijove_34_Composite.png/Ganymede Perijove 34 Autor NASA/JPL-Caltech/SwRI/MSSS/KevinM.Gill
https://upload.wikimedia.org/wikipedia/commons/c/c2/Miranda_mosaic_in_color_-_Voyager_2.png https://www.flickr.com/photos/1970 38812@N04/53467048107/Autor zelario12
https://upload.wikimedia.org/wikipedia/commons/thumb/b/b1/Uranus_Montage.jpg/http://solarsystem.nasa.gov/multimedia/display.cfm?Category=Planets&IM_ID=10767
http://solarsystem.nasa.gov/multimedia/gallery/Uranus_Montage.jpg Author NASA/JPL
https://upload.wikimedia.org/wikipedia/commons/thumb/4/4e/PIA00039_Titania.jpg/http://ciclops.org/view/3651/Titania_-_Highest_Resolution_Voyager_Picture Autor NASA/JPL
https://upload.wikimedia.org/wikipedia/commons/thumb/2/2e/Apollo_15_Lunar_Rover_and_Irwin.jpg/http://www.hq.nasa.gov/alsj/a15/images15.html Autor NASA/David Scott
https://commons.m.wikimedia.org/wiki/File:Solar_System_true_color.jpgCactiStaccingCrane
https://upload.wikimedia.org/wikipedia/commons/thumb/d/d5/Comet_McNaught_at_Paranal.jpg/jpghttp://www.eso.org/public/images/mc_naught34/Author ESO/Sebastian Deiries European Southern Observatory (ESO).
https://upload.wikimedia.org/wikipedia/commons/thumb/d/d7/Terrestrial_planet_sizes_3.jpg/Orbiter Mission (30055660701).png (ISRO / ISSDC /Justin Cowart)Author CactiStaccingCrane
https://upload.wikimedia.org/wikipedia/commons/thumb/6/67/Planet_collage_to_scale_(captioned).jpg/User:MotloAstro(Sun); NASA Author CactiStaccingCrane
https://upload.wikimedia.org/wikipedia/commons/thumb/2/2d/The_Mysterious_Case_of_the_Disappearing_Dust.jpg/NASA/JPL-Caltech
https://upload.wikimedia.org/wikipedia/commons/thumb/e/e3/Magnificent_CME_Erupts_on_the_Sun_-_August_31.jpg/Flickr : Magnificent CME Erupts on the Sun - August 31Autor NASA Goddard Space Flight Center
https://upload.wikimedia.org/wikipedia/commons/thumb/a/ae/Phoebe_cassini_full.jpg/JPL image PIA06064 Author NASA/ JPL/Space Science Institute
https://upload.wikimedia.org/wikipedia/commons/thumb/3/3a/Mare_Imbrium-AS17-M-2444.jpg http://nssdc.gsfc.nasa.gov/imgcat/html/object_page/a17_m_2444.html
http://www.lpi.usra.edu/resources/apollo/frame/?AS17-M-2444Autor NASA
https://upload.wikimedia.org/wikipedia/commons/a/a6/Moon_phases_00.jpg Orion 8
https://upload.wikimedia.org/wikipedia/commons/thumb/8/81/Artemis_program_hls-ascending.jpg/https://www.nasa.gov/feature/nasa-seeks-input-from-us-industry-on-artemis-lander-development Autor NASA
https://upload.wikimedia.org/wikipedia/commons/3/3e/Deep_Impact_HRI.jpegNASA/JPL-C

THE SOLAR SYSTEM: THE SUN AND THE PLANETS

altech/UMDhttp://discovery.nasa.gov/images/67_secs_after_impact.jpg archive copy at the Wayback Machine
https://upload.wikimedia.org/wikipedia/commons/thumb/c/c4/ALH84001.jpg/http://www-curator.jsc.nasa.gov/curator/antmet/marsmets/alh84001/ALH84001,0.htmAutor NASA
https://upload.wikimedia.org/wikipedia/commons/thumb/1/17/PIA22083-Ceres-DwarfPlanet-GravityMapping-20171026.gif/https://photojournal.jpl.nasa.gov/archive/PIA22083.gifAuthor NASA/JPL-Caltech/UCLA/MPS/DLR/IDA
https://es.m.wikipedia.org/wiki/Archivo:Vesta_full_mosaic.jpg View of Vesta Autor NASA/JPL-Caltech/UCAL/MPS/DLR/IDA
https://upload.wikimedia.org/wikipedia/commons/7/72/Iau_dozen.jpg (IAU/NASA) Martin Kornmesser NASA/ESA and the Hubble Heritage Team"
https://upload.wikimedia.org/wikipedia/commons/thumb/8/84/The_Four_Largest_Asteroids_(unlabeled).jpg/Ceres and Vesta images: NASA/JPL-Caltech/UCLA/MPS/ DLR/IDA Pallas image: NASA Hygiea image: Astronomical Institute of the Charles University: JosefĎurech, Vojtěch Sidorin Image modified by PlanetUser.
https://upload.wikimedia.org/wikipedia/commons/8/86/The_Four_Largest_Asteroids.jpg Ceres and Vesta images: NASA/JPL- Caltech/UCLA/ MPS/DLR/ IDA Pallas and images: ESO Images compiled by PlanetUser and by kwamikagami
https://upload.wikimedia.org/wikipedia/commons/f/ff/Nereid_-_Simulated_View.jpgPlanetUser
https://upload.wikimedia.org/wikipedia/commons/thumb/4/47/Moons_of_Saturn_-_Infographic_(15628203777).jpg/Kevin Gill from Nashua, NH, United States
https://upload.wikimedia.org/wikipedia/commons/thumb/8/82/Enceladus_Cross-section.jpg/https://www.flickr.com/photos/5078505 4@N03/36403387400/Author NASA-GSFC/SVS,NASA/JPL-Caltech/Southwest Research Institute
https://upload.wikimedia.org/wikipedia/commons/thumb/4/41/Enceladus_(14432622899).jpg/Kevin M.Gill Flickr set Hourly Cosmos
https://upload.wikimedia.org/wikipedia/commons/thumb/4/4d/PIA21913-DwarfPlanetCeres-OccatorCrater-SimulatedPerspective-20171212.jpg/NASA/JPL-Caltech/UCLA/MPS/DLR/IDA Ander weergawes Oblique view of crater
https://upload.wikimedia.org/wikipedia/commons/6/6d/Oberon_in_true_color_by_Kevin_M._Gill.jpghttps://www.flickr.com/photos/kevinmgill/50906003243/Author Kevin M.Gill

www.ingramcontent.com/pod-product-compliance
Lightning Source LLC
Chambersburg PA
CBHW071955210526
45479CB00003B/946